To: Colin
From: Grandpa, Grandma & "Big Guy Erik"
Christmas 1988

Lookout Land

Published by Modern Publishing,
a Division of Unisystems, Inc.

Printed in Singapore.

The Little Lookouts™
Discover Numbers

Written By Rosemary Grimm
Illustrated by Nathan Y. Jarvis and George F. Martin

Modern Publishing
A Division of Unisystems, Inc.
New York, New York 10022

1
one

"Tomorrow is my birthday," declared Amelia. She circled **one** date on her calendar with her **one** crayon. "I want to have **one** big birthday party. I'll invite all of the Little Lookouts: Christopher, Joan, Lewis, Clark, and Cortez."

2
two

"I'll help you plan your birthday party," said Joan. "Let's make **two** lists of the things we need." Joan brought out **two** pencils and **two** sheets of paper.

"I found these **two** party books to help us," said Amelia. The **two** Little Lookouts then began to plan Amelia's birthday party.

3
three

Amelia and Joan thought about the birthday party for a long time. At **three** o'clock, Lewis arrived. He wanted to show Joan and Amelia his **three** fish.

4 four

At **four** o'clock, Clark arrived. He wanted to help.

"I think all the guests should eat **four** huge bananas," he said.

"Okay, but we'll make sure everyone has **four** napkins, too," said Joan.

After Lewis and Clark left, Joan said, "Now let's make **five** party invitations."

5
five

Joan and Amelia used **five** bright crayons to color the **five** invitations. They made **five** envelopes and put **five** stickers on them.

The next day, Amelia and Joan decorated the table for the birthday party. Joan put **six** plates and **six** forks on the table. Amelia put **six** cups on the table.

6
six

"Now there are **six** places to sit," said Amelia. "One for each Little Lookout."

“The table will look pretty with these **seven** flowers and **seven** vases,” said Joan. “I like the **seven** bright colors.”

7
seven

"I like these **seven** bright lights," said Amelia.

“I found **eight** party hats,” laughed Amelia. “They are all funny-looking.”

8
eight

One party hat had **eight** polka dots on it. One hat had **eight** stripes. Some even had **eight** streamers coming up out of the top.

Joan used **nine** bowls to make **nine** cupcakes for the birthday party. She decorated the cupcakes with **nine** cherries.

9
nine

Next she baked **nine** cookies and **nine** brownies.
"I hope I've made enough," she worried.

Amelia went to her room and changed into her party clothes. Amelia's room was decorated with **ten** Little Lookout flags. On her bed were **ten** bows and **ten** bracelets.

“My guests will be here soon,” said Amelia. “I’d better decide which bow to wear in my hair and which bracelet to put on.”

1

2

3

5

4

Amelia is ready for her birthday party. Count all the objects before her guests arrive.

6

7

8

9

10

1
one

Finally, it was time for Amelia's birthday party. Joan put **one** birthday cake on the table. Amelia waited by the Little Lookouts' **one** large tree for the guests to arrive.

Cortez and Christopher were the first **two** guests to arrive. Cortez brought Amelia **two** boxes of candy and showed Amelia and Joan his **two** new shoes.

"Those shoes are pretty," said Joan. "Do you like my **two** new flowers?"

Lewis was the next guest to arrive. He brought **three** presents for Amelia.

3
three

Amelia clapped for joy. "Thank you, thank you, thank you," she said **three** times.

"I wonder where Clark is," said Joan.

"Here I am," cried Clark. "Look what I've got for you — **four** balloons!" Clark carried the **four** balloons proudly.

Each balloon had **four** shapes on it — **four** squares, **four** triangles, **four** circles, and **four** rectangles.

4
four

"Those balloons are great decorations for the party," said Amelia. "Thank you for bringing them."

"Happy Birthday, Amelia!" shouted the Little Lookouts. Amelia's birthday cake was decorated with **five** roses and **five** candles.

5

five

Amelia blew out the candles. “The cake looks delicious,” she said.

Lewis had **six** straws. He gave one to each guest. Then Joan poured **six** glasses of punch.

6
six

"Is it time to eat the birthday cake, yet?" Amelia asked.

Christopher wanted to play **seven** songs at the party. He picked out **seven** records.

7
seven

“Can we listen to the Monkey Shuffle?” asked Clark.

“Now for some entertainment,” said Clark. He juggled **eight** rocks. Then he juggled **eight** bananas. Then he juggled **eight** spoons.

"Can we eat now?" asked Amelia impatiently.

Joan brought out her **nine** cupcakes with the **nine** cherries on top. Cortez brought out the **nine** cookies.

9
nine

"I've cut the cake. Let's eat," said Clark.
"Your party is fun, Amelia," said Lewis.
"More punch, please," munched Amelia.

After the Little Lookouts ate, they played party games. They bobbed for **ten** apples. They played pin-the-flag-on-the-ship and used **ten** flags.

10
ten

"We had a great time at your party, Amelia," said Christopher.

"Great," said Amelia. "I've already started planning next year's party." The Little Lookouts just laughed.

1

2

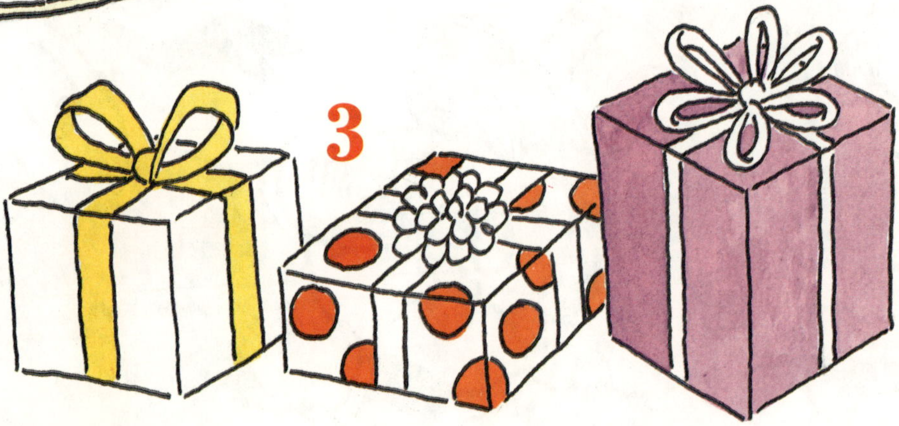

3

4

5

Amelia had a wonderful birthday party with all her Little Lookout friends. Count all the fun things she had at her party.

6

7

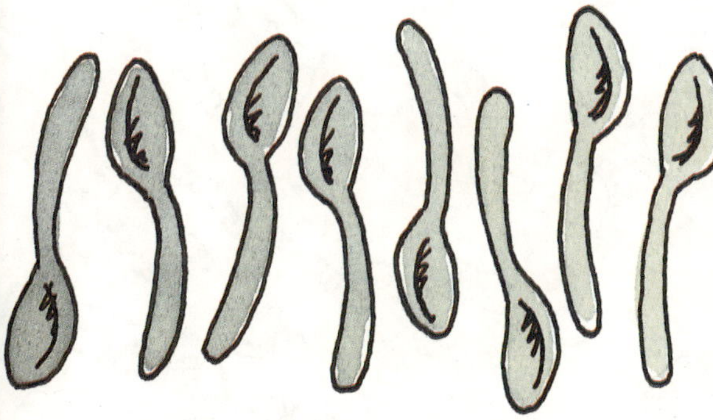

8

9

10

OCEAN
WITCH WAY FOREST
BALONEY SANDY WITCH'S CASTLE
FORGET-ME-NOT POND
Very Cranberry Canal
OCEAN